EUROPE
ASIA
AFRICA
INDIAN OCEAN
PACIFIC OCEAN
OCEANIA
ANTARCTICA
11
12
13
10
14
9
15
16
17
7
8

Titles in this series

Animal Alphabet Book
Birds
Plants
Animals
Animal Homes

The publishers would like to thank the staff of World Wildlife Fund for their help in making these books.

Acknowledgment:
Front and back cover and endpaper illustrations by Stephen Lings.

LADYBIRD BOOKS, INC.
Auburn, Maine 04210 U.S.A.

Printed in England

WORLD WILDLIFE FUND

Animal Homes

written by GILLIAN DORFMAN
illustrated by COLIN WOOLF

Ladybird Books
Produced in association with World Wildlife Fund

This animal is a giraffe. He does not live alone.
He lives with other giraffes, and with other animals.

He lives with plants, too. All these living things – both plants and animals – live together.

This is the giraffe's home. It is made up of lots of things – rocks and water, air, trees, and grass. These give the grassland its shape and color.

Sometimes rain falls and makes the giraffe's grassland home wet.

Rain, sun, and wind make up the weather in this grassland.

The heat from the sun dries the land.

Sometimes the wind blows. The wind stirs the grass and the trees.

The giraffe likes living here. Everything is just right for him in his grassland home.

The giraffe is tall, so he can eat the tasty leaves in the treetops, which none of the other animals can reach.

The land is flat, so the giraffe can watch out for his enemy, the lion.

He cannot live in the cold place where these penguins live.

No trees grow here, so there are no green leaves for the giraffe to eat.

Every animal has its own special home. Let's see where some other animals live.

Mountains are high, steep places. On the mountaintops, where the air is cold and the wind blows hard, there is only rock and ice and snow.

But on the lower slopes, forests grow.
The giant panda lives here.

The giant panda loves to eat the bamboo that grows on the mountain slopes.

Other animals live in the mountains, too.

The snow leopard, with its pale fur coat, can stalk and surprise its prey, the bharal sheep.

The golden eagle flies low over the mountain, looking for small animals to swoop down on.

The jumping spider feeds on small flies and other insects.

The mountain brook has fresh, cool water. It babbles and bubbles as it rushes down the hillside. This is the home of the trout.

The trout lays her eggs in the shallow waters of the brook.

The brook becomes a wide river that flows slowly and silently to the sea. The river is home to the kingfisher.

The kingfisher is very good at catching fish. He catches them with his long, sharp beak as he dives into the water.

The pike lurks in the water weeds, ready to lunge at smaller fish.

The pine tree grows in cold lands. It has leaves all year round. The leaves are like needles, and the pine cones hold seeds. The porcupine lives in this dark forest of pine trees.

The crossbill's beak is specially shaped so that he can force pine cones open to eat the seeds.

The porcupine has quills as sharp as arrows, which help protect her when wolves and bears attack. When the snows come, she spends much of her time in the pine trees, gnawing bark.

Other animals live here, too, and some just come to visit for a short time.

In the cold months the caribou lives in the forest and feeds on tree twigs.

Pine needles are not tasty, but the red squirrel finds the seeds from the pine cones good to eat.

These woodland animals live in a place that is warm in summer and cold in winter. Let's see how they live when the world around them changes.

In summer, the fox sheds his heavy winter fur.

These wood warblers have come from far away to find food and to have babies.

As the weather gets colder, the wood warblers fly away to warmer places where there is plenty of food to eat.

The hedgehog hides away and sleeps until the weather gets warm again.

The squirrel and the jay eat food that they stored before the winter set in.

This is the jungle, where it is always hot and wet, and the plants grow big and fast. Tall trees reach for the sun. Ferns grow in the shade, and there are plants that perch on, or climb up, or twist around other plants.

At night the jaguar hunts alone. He has a spotted coat that helps him hide in the shadows and creep up on his prey.

The anaconda can grow up to 35 feet long. He smothers his prey by coiling around it.

The jungle is home to millions of creatures.

The harpy eagle patrols the sky, looking for monkeys to swoop down on.

The tiny hummingbird hovers in the air as he sips sweet juices from the flowers.

The desert is a hot, dry place. Few plants can grow in the desert, but camels can live here.

The camel stores fat in his hump. He uses this spare food when there are no plants to eat or water to drink.

Wide, padded feet keep the camel from sinking into the sand.

Other animals live here, too.

The sand grouse flies far to find water. It soaks its chest feathers to carry water back to its young.

The tiny fennec fox has huge ears to help him hear his prey when he hunts at night.

The jerboa shelters from the sun. He comes out at night to feed.

Many animals live in the salty waters of the ocean. Some swim and drift in sunlit waters, and others live below in the cold, dark depths.

As it drifts, the jellyfish catches food that crosses its path.

This school of fish eats tiny plants.

Sharks swim fast when they are hunting.

Animals that live at the top of the ocean cannot live at the bottom, and those that live at the bottom cannot live at the top.

Some creatures have big mouths to catch food that sinks to the ocean floor.

Look at the angler fish's light. Perhaps it is her way of attracting her prey.

In the far north there is a frozen sea. It is a cold and windy place. Summers are short, and winters are long and dark. No plants can live on the snow and ice, but the polar bear lives here.

Polar bears are born in the winter, in a den beneath the snow.

The polar bear's feet can grip the ice so that he does not slip and slide.

Other animals live here, too.

The walrus and the seal each have a layer of fat beneath their skin to protect them from the cold.

Every animal has its own special place. There, it fits in with its surroundings, and with plants and other animals. This is its home.

WWF ©

Many of our world's plants and animals are in danger. People have destroyed or polluted the places in which they live or grow. Some animals have been hunted until every one of them has been killed. This is what happened to the dodo, an amazing flightless bird that once lived in Mauritius. The same thing could happen to gorillas, tigers, and whales unless we do something to save them now.

WWF (World Wildlife Fund) was set up to warn people about the dangers threatening the earth's wildlife. If we know and care about what happens to our world, we may be able to protect it before too much damage is done.